Grace Porter Hibbard

Wild Poppies

Grace Porter Hibbard

Wild Poppies

ISBN/EAN: 9783744651813

Printed in Europe, USA, Canada, Australia, Japan

Cover: Foto ©berggeist007 / pixelio.de

More available books at **www.hansebooks.com**

WILD POPPIES

BY
GRACE HIBBARD

BUFFALO
CHARLES WELLS MOULTON
1893

Copyright, 1893,
By GRACE HIBBARD.

Printed by C. W. Moulton, Buffalo, N. Y.

DEDICATION.

TO the Military Order of the Loyal Legion of the United States, and more especially to California Commandery, this book is respectfully dedicated.

INDEX.

	PAGE
The Spirit of the Spring	1
Dreaming	4
Bells of Venice	5
Blue Bells	6
My Father's Sword	6
Under the Pines	7
The Old Slave's Lament	9
Fishing	10
All Souls' Eve	10
Apple Blossoms	12
The Soldier's Son	13
Up from the Sea	14
"Night's Candles are Burned Out"	15
My Playmate	15
Love's Immortality	17
How shall Jeanne de'Arc be Painted	20
"Je Te Rejoins"	21
An April Snowflake	22
The Old War Horse	22
Hope	24
Wild Roses	25
Snow on the Plains	27
Away from Home	28

Index

	PAGE
Poem	28
Found	30
The Crystal Bells of Santa Helena	31
Ancestor Mine	32
The Astronomer's Wife	32
The Pagan Girl's Prayer to the Sun	34
A Vignette	34
Waiting for Colin	35
Under the Orange Trees	36
The Old "Hartford"	37
Shadow Land	38
Wild Poppies	39
The King's Return	40
Someone	42
Cast Aside	43
Memorial Day	43
A Legend of Calvary	44
A White Chrysanthemum	45
The Heart's Firelight	45
Discovery of the Sunshine Mine	46
A Dream	50
My Mother's Birthday	51
A Winter's Day	51
The Sea is a Grave To-day	53
Morning Glories	54
One Wild March Night	55
A Prince	56
Away	56

Index.

	PAGE
Wild Violets	57
The Royal Succession	58
On the Beach	59
A Valentine	60
New Year's Fancies	60
Either Way	61
My New England Home	62
The Loveliest Picture	63
A Prayer	64
In the Gloaming	65
Compensation	66
A Cigarette	66
Waiting at the Gate	67
In the Cathedral	68
San Juan by the Sea	69
Before the Holidays	69
After the Holidays	71
Whither	72
Suspense	72
Safe	73
Sunset Fancies	74
Where	74
To Ada Rehan's Picture	75
Alpine Barry	76
To———	78
The Sapphire Sea	79
A Paris Bonnet	80
All that Remains	82

Index.

	PAGE
The Sun has Gone Down	83
Do They Know	84
California	85
My Watch	86
Night at Sea	88
A Laurel Wreath	88
Day Dreams	89
Eternal Silence	90
To-morrow will be Bright	91
Under the Sentence of Death	92
The King's Daughter	92
Changed	94
Farewell	95
Pictures on the Wall	96
Beside the Sea	99
A Golden Pathway	99
New Year's Eve	100
My Traveler	101
Every Morning	101
Jewels from Under the Sea	102
Too Soon	103
Platonic Friendship	104
Under a Mimosa Tree	105
Two Stars	105
The Clock on the Tower	109

WILD POPPIES.

THE SPIRIT OF THE SPRING.

WE made our home in the wilderness,
  The wilderness of billowy grass,
That rose and fell at the tide of winds,
  But lay at noontide a sea of glass.

I was an artist, who sought to catch
  The sunset's glory on prairie wide;
A picture to paint was my fond hope,
  For the Salon—and she was my bride.

Before our cabin a cottonwood grew,
  Whose heart-shaped leaves, like humming-
    bird's wing,
Fluttered, and quivered, on slender stems,
  And in its shadow a bubbling spring.

Summer had passed like a spirit by,
  The cottonwood's leaves were sere and gray,
And the corn-stalks stood like sentinels,
  Summer's outposts, that sad autumn day.

But alas! the sunset I had sought
  To capture on canvas, for the Salon,
Still burned in the sky, and in my brain,
  And the radiant summer was gone.

Wild Poppies.

The noon was hot, and breathless, and still,
   The white clouds rose like mountains high,
Peak above peak, grim giants at war,
   In the far away, blue, western sky.

I mounted my horse, that sultry noon,
   Not heeding her voice who bade me stay
Nor the mute appeal of her white arms,
   Held out to me as I rode away.

I rode, and rode, for many a mile,
   My sombrero down over my eyes,
And smoked cigarettes, and cursed my fate,
   'Till a tint of gray crept o'er the sky.

Was my brain maddened, or did I hear
   The whisper of demon from below?
"There'll be no red in the sunset to-night;
   Paint thou the prairie fire's red glow."

The air was breathless, and still and hot,
   The billowy grass a motionless sea,
No breeze was coming from east or west;
   I threw my cigarette far from me.

A torch of fire my cigarette;
   The dry grass changed to fluttering wings
Of scarlet and gold, then serpents crawled
   In sinuous paths, like living things.

The Spirit of the Spring.

Wild with delight at the deed I had done,
   I'd not taken thought; was mine the blame
That like a demon out of the west
   On wings of blackness, the wild winds came?

I thought of Pharaoh's struggling hosts,
   As frantic I crossed the fiery sea,
To rescue her, far dearer than life,
   And some way a path was made for me.

For she was alone, my darling one;
   In the fire's path my cabin stood;
I saw, like shower of falling stars,
   The blood-red leaves of the cottonwood.

Before our ruined cabin I stood,
   Wild with despair; 'neath the leafless tree,
Calling my darling's name o'er and o'er,
   Begging my darling to come back to me.

Up out from the spring my darling came,
   A look of ecstasy on her face:
My picture, "The Spirit of the Spring,"
   In the Paris Salon, had a place.

DREAMING.

IDLY sitting by my window, fair dreams dreaming—
Dreaming snowy clouds are castles seeming,
Built on gray rocks in the sky sea lying,
Stormed by golden sunbeam arrows flying.

Idly sitting by my window, fair dreams dreaming—
Dreaming snowy clouds are white waves gleaming,
On the tropic blue of sky sea dashing,
In the brightness of the sunset flashing.

Idly sitting by my window, fair dreams dreaming—
Dreaming white clouds are cherub faces beaming,
With bright, fleecy hair around them streaming.
In the twilight idly sit I dreaming.

Idly sitting by my window, fair dreams dreaming—
Castles proud, white waves, cherub faces beaming,
Turned to empty air, like all earth's dreaming;
But above me, lo! the stars are gleaming.

BELLS OF VENICE.

SILENCE o'er city fair,
    Not a breeze sighing,
Silence in palace old,
    At the day's dying.

Gold in the sunset sky,
    And on sea lying,
Long lines of golden light,
    Like arrows flying.

Boats on the paths of blue,
    Blue skies o'er bending,
Silence at sunset's hour,
    At the day's ending.

When lo! the many bells,
    From each church tower,
Ring out in melody,
    At sunset's hour.

Silence unbroken save
    For sweet bells ringing,
As through the sunset's gate
    Day's flight is winging.

BLUEBELLS.

THE unseen fingers of the air
    Set all the bluebells ringing.
My thoughts like birds that homeward fly,
    Across the sea went winging.

To "banks and braes" where bluebells grow,
    'Neath trees where birds are singing,
Their home and mine—did others hear
    The bonnie bluebells ringing?

MY FATHER'S SWORD.

CLOSE sheathed in its scabbard on the wall,
    Hangs, draped with faded, crimson, silken
        sash,
    My father's sword.   No longer sabres flash,
Nor cannon flame and blaze from fortress tall.
Sweet peace, like snowy dove, for many years,
    Has hovered o'er our land, a spirit fair;
    But in a single night my mother's hair

Was changed to white; my face baptized by tears.
For in the sunny south one springtime day,
  A soldier fell, his sword clasped in his hand.
The wild birds sang as blithe as e'er before,
And apple blossoms crowned the joyous May.
  Upon a home in the far northern land,
  A shadow fell, that lifted nevermore.

UNDER THE PINES.

BEFORE the grate in the firelight,
  On the night when the year grows old,
Watching the smoke curl phantom-like,
  And the coals turn to living gold,

I sit and dream as I listen
  To sweet clamor of New Year's chimes,
And whisper low the vows I made
  In the moonlight, under the pines.

I have left the dazzling ball-room;
  Decked in jewels that brightly gleam,
In my dress of pearl-white satin,
  I have come to my room to dream,

I have left music and dancing,
   The soft, perfumed, tropical air,
The eyes and the voices that told me;
   "The Rose of the Mountains, is fair."

Once more I am Mabel; daughter
   Of "Old Ben" of the "Blue Bird Claim;"
I hear my boy lover asking:
   "Wild Rose, will you love me the same

When you go with your father's sister
   To the city so far away?
Will my "Blue Bird" of the mountains
   Come back to the home nest, some day?"

Upon our sure-footed ponies,
   Up the zigzag cañon wild,
We had wandered to gather flowers,
   In the twilight of springtime mild.

The giant peaks in the gloaming
   Seemed touching the shining stars;
The moonlight upon the pine trees
   Turned their branches to golden bars.

I answered, with hand uplifted,
   "Just as long as the North Star shines,
I will keep the vows I made you
   In the moonlight, under the pines.

So I've left the dazzling ball-room,
  Decked in jewels that brightly gleam,
In my dress of pearl-white satin,
  I have come to my room to dream.

I kneel in the glowing firelight,
  As I listen to New Year's chimes,
And whisper low the vows I made,
  In the moonlight, under the pines.

THE OLD SLAVE'S LAMENT.

THAR was singin', thar was dancin'
  In de little cabins, long ago;
An' cotton growin' in de fields
  As white as northern snow.
In Massa's house lights twinkled,
  And de young folks danced—ho! ho!
Reckon de likes ob dose good times
  Pore ole Pete will neber know.

'Spec de birds do all de singin',
  An' de sunshine all de dancin' on de floor:
An' de lights go twinkle, twinkle,
  In Massa's house no more.

Ole Pete is sometimes hungry,
  But he'll let the chilluns know,
Thar was singin', thar was dancin'
  In de cabins long ago.

FISHING.

THE moonlight cold and still
  In net-like golden bars,
Lies on the waters blue
  To catch reflected stars.

ALL SOULS' EVE.

I AM all alone in my room to-night;
  It is "All Souls' Eve" when they say the dead
For a single night can walk the earth
  And then go back to their lone church-yard bed.

Outside of the house the autumn winds blow;
  (Do I hear the sound of the garden gate?)
I have decked the room with flowers they loved,
  And placed a warm mat before the bright grate.

All Souls' Eve.

Down memory's pathway they come to me;
   My soldier father, and close by his side,
My golden-haired mother, who left her child,
   When the cruel word came that he had died.

With only Carlo, my St. Bernard friend,
   I was left alone in the cold, world wide;
My dog was sent to the holy monks
   To save men lost on the bleak mountain's side.

I knelt sad before the crucifix white,
   And cried; "O Mother, I am all alone!
There is no one to love me; let me go
   To-night with you to your heavenly home."

I heard the sound of the garden gate, and
   "Bernadine, Bernadine listen to me,
I, Victor, swear true by the holy dead,
   Of all the wide world I love but thee."

APPLE BLOSSOMS.

SHE gave him an apple blossom
   One day in the sweet springtime,
She did not know its meaning,
   That it whispered "My heart is thine."

But someway her love had wandered
   Away from one stern and cold,
As dainty, pink-white blossoms
   Drift away from apple trees old.

And he read the old sweet story
   In glance of her blue eyes meek,
And pink of apple blossoms
   As it flitted across her cheek.

THE SOLDIER'S SON.

<small>Read at a banquet of California Comandery, Loyal Legion, by Col. W. R. Smedberg, Recorder.</small>

IN the sunset's glory they stand
    Together, the heroes in blue;
The slanting sunbeams rest on their arms,
    And the mystic river's in view.

To the other side their comrades
    Have crossed at the word of command,
And brighter far than earth's laurel wreaths
    Are the crowns of that martyr band.

There is one who died long ago,
    Who for freedom his young life gave;
Each springtime by loving hands are placed
    Fair flowers on that soldier's grave.

In a shadowed home, on the wall,
    Hangs the sword he wielded so well;
His gold-barred shoulder-straps are kept
    That he wore on the day he fell.

To-night his son at your hands, craves
    The cross to be placed on his breast,
The badge that his father's valor won,
    The soldier long gone to his rest.

UP FROM THE SEA.

WRAPPED in the cold, silver mist so white,
    Up from the sea come the silent dead;
Through streets of the city with noiseless tread,
They wander together—'tis All Souls' Night.
One looks in the window, where long ago
    Beloved at the hearthstone she had a place,
    And she gazes long at a manly face.
"I love you, my husband," she murmurs low.
Men shuddering hurry along the street;
    They shiver at touch of the cold white mist,
    And they long for the morning's warm sunlight;
They know not 'tis spirit they love they meet,
    They feel a horror they cannot resist,
    Forgetting, alas! it is All Souls' Night.

"NIGHT'S CANDLES ARE BURNED OUT."

LAST night when stars were lighted one by one,
 Eyes blue as summer skies,
And bright like stars that shine—
 Dear dying eyes—
Looked into mine.

* * * * * *

"Night's candles are burned out;" the day is here;
 The radiant blue eyes
So bright, like stars that shone—
 See fairer skies;
I am alone.

MY PLAYMATE.

"I WILL come on a coal black horse,
 I will come in ten years, Fay,
When the apple blossoms are pink and white,
 In the merrie month of May."

'Twas my little playmate who spoke.
  I was eight years old that day.
We stood in the orchard under the trees;
  He was soon going away.

Far away from the sea-swept coast,
  Far beyond mountains and plains,
To where rivers rolled over sands of gold,
  And mountains had golden veins.

He said: "To the sunset I'll ride;
  I shall never lose my way;
Remember and watch when the apple trees bloom,
  In the merrie month of May.

When my playmate left me for school,
  From his small blouse, blue and white,
He brushed away just a few boyish tears,
  Then he vanished from my sight.

*   *   *   *   *   *

In front of our cabin I stand,
  My home on the mountain-side,
In one hand are blossoms of wild plum trees
  From the cañon, deep and wide.

With my other hand I now shade
  My eyes from the eastern sun,
And look for a rider on a black horse,
  I'm sure my playmate will come.

LOVE'S IMMORTALITY.

IN far-off classic land,
   Blazing torch in her hand,
  On a high tower,
Stood Hero, young and fair,
With halo of bright hair,
  At the midnight hour.

Out on the inky night
Fluttered the red torchlight,
  To guide her lover;
Flaring in the keen blast,
Then lost, like star o'ercast,
  Held high above her.

Not half a year ago
In vestal robes, like snow,
  To sound of lyres,
Upon the altar bright
On Venus' festal night,
  She fed the fires.

Child of a noble Greek,
With face of virgin meek,
  Eyes of heaven's blue;
Mid clouds of incense rare,
She stood, a priestess fair,
  To the Goddess true.

Wild Poppies.

Love made her vows as naught,
Sweet lesson she was taught
   In one short hour.
Dark eyes of Thracian youth
Told her the wonderous truth
   Of love's grand power.

Banished to island lone
To castle ivy grown,
   Alone they left her.
Love can bridge waters wide,
So, soon to Hero's side
   Came young Leander.

Swimming the Helespont
Nightly became his wont
   To Hero's tower.
First by the full-moon's light,
Making a pathway bright
   At moon-rise hour.

But came a stormy night
With lightnings flashing bright
   And sad winds wailing.
Moonless and starless sky,
Black clouds o'er gray sky fly;
   Pirate ships sailing.

Love's Immortality.

Love can make darkness light;
Out on the stormy night
  Hero's torch flashes.
Leander sees the gleam
And in the angry stream
  Heedlessly dashes.

Pitiless breakers roar
Louder than e'er before
  Seem to the swimmer.
Darker the gray sky grows,
Wilder the storm wind blows;
  Hero's light dimmer.

She from her tower prays
Goddess of her young days
  To save her lover.
Brighter the lightnings flash,
Louder the breakers dash;
  No stars above her.

Down on the rocks below,
Mid breakers white as snow,
  There he lies dying.
Down to his side she leaps,
Torch in her hand she keeps;
  Meteor flying.

Long line of golden light,
Lighting fair Hero's flight,
  Through death's dark portal.
Such love that does not shrink,
Even from death's dread brink,
  Must be immortal.

HOW SHALL JEANNE D'ARC BE PAINTED?

AS child shall she be painted, watching her father's
      flock's,
  Wandering among the lambs, the gentlest there;
Green summer fields, wild-flowers, a few tall trees,
  A crown of golden buttercups upon her hair?

Or shall Jeanne be painted as a warrior in armor
  Leading to battle soldiers, though but maiden
      fair;
Riding on plunging war-horse, a lone guiding star,
  Helmet in place of buttercups upon her hair?

Or in the market-place of Rouen shall he paint her,
  Bound to a stake, with cruel chains, her life work
      done;
Faggots and tiny wings of fire about her
  Crowned with a halo by the golden setting sun?

"*Je Te Rejoins.*"

No. Let the artist paint her as she listens—
Listens to whisperings from the far heavens blue;
Voices unheard save by the Maid of Orleans,
Telling Jeanne d'Arc the mighty work she has to do.

"JE TE REJOINS."

Suggested by the suicide of an aged French florist, upon the grave of his wife.

" I CANNOT live without thee,"
Were the words I whispered to thee, dear one,
In our bright sunny France, long years ago,
When we were young, and thou wert fair to me.
Sweetheart, I loved thee so,
And now, "I go to thee."

"I cannot live without thee."
Bright flowers I have laid upon thy grave
For many and many a dreary day.
Without thee life has no more charm for me;
Bereft I cannot stay,
And so "I go to thee."

"I cannot live without thee."
The key of death I hold within my hand;
Alone beside thy grave in church-yard drear,
God pity, pardon if I use the key;
Earth vanishes away;
And thus, "I go to thee."

AN APRIL SNOW-FLAKE.

THE apple blossoms held pink-white cups
    To catch the April shower;
When from the sky came floating down
A tiny crystal flower.
It was only a little snow-flake white,
But the sun peeped out from behind a cloud
And it turned to a jewel bright;
In another moment the jewel bright
Was changed to a tear in the flower cup white.

THE OLD WAR HORSE.

OH! never again to march at the head of a column;
Only to graze in the field at the edge of woods solemn

Only to drink at the moss-covered trough in the meadow;
Only to stand in the sunshine and then in the shadow.

The Old War Horse.

Aye! once more to march at the head of a warlike column,
Leading veterans and soldiers marching to music solemn.

Soldiers marching to decorate graves, carrying flowers,
Passed fields where "Old Joe" roamed alone through long summer hours.

Back turned the ears of the war horse at martial strains sounding,
Forward his ears, a step, then into the dusty road bounding,

Pawing the air, and then wheeling, and leading the column,
Tears starting to eyes and hats raised at a sight so solemn.

Tears to the memory of their Colonel, young, loved and brave;
Whom "Old Joe" bore long ago to battle, and to his grave.

HOPE.

OUT on the sea, out on the sea in a storm;
　　Lightning flashing and cutting like swords the
　　　　inky black sky;
Down in the trough of the sea, then lifted by waves
　　　　mountains high,
Rides a ship, alone in the tempest; destruction is nigh.
"Down with the anchor," the Captain's cry.

Like a tired bird, a bird with its wings at rest,
Swinging at anchor on watery waste, 'neath bright-
　　　　'ning sky,
The stately ship rides; like flags of gold on the three
　　　　masts high
Sunbeams gaily flutter; winds whisper, "The haven
　　　　is nigh."
"Anchored steadfast and sure," the glad cry.

Out on the sea, on the troubled sea of life;
Lightning flashing and cutting like swords the inky
　　　　black sky;
Down in the depths of woe, lashed by the cruel waves
　　　　so high,
Drifting along, gladness so far away, despair so nigh;
"Hope the soul's anchor," our Captain's cry.

WILD ROSES.

TO-night before the bright foot-lights,
    Decked with jewels that flash and gleam,
In robe of velvet and ermine,
    I played the part of a queen.

Far upward my voice soared bird-like,
    'Till it seemed to reach the blue sky,
Then changed to notes, low and plaintive,
    Like the soft summer's wind low sigh.

Before me were beautiful women,
    The courtly, the stately, the grand;
There were men of wealth and fashion
    Who had begged me for my hand.

At my feet fell fairest of flowers
    That perfumed the tropical air,
In one was hidden a jewel
    That shone in the gaslight's bright glare.

Some one tossed a few wild roses,
    But little the dazzling crowd guessed
Why I left the others unnoticed
    And fondly clasped them to my breast.

Again I was poor little Inez,
　　The fisherman's child by the sea;
The cluster of wild pink roses
　　Brought a moonlight picture to me.

The round moon upon the waters
　　Made a pathway of golden light,
Across it a ship was sailing—
　　I was watching it out of sight.

In that brave ship my boy lover
　　Sailed away out into the night;
I held in my hand wild roses,
　　As I watched it vanish from sight.

To-night, not knowing, not dreaming,
　　I sang to one, just home from sea;
'Twas the hand of my boy lover
　　Tossed the sweet wild roses to me.

SNOW ON THE PLAINS.

LAST night across the glory of the sky, purple clouds lay;
The gray-brown, arid plains wandered away and met the sunset bright.
Like rusted blades the lush grass rustled in the balmy air.
The sage-brush in the gloaming seemed like timid deer in flight,
Or Indians, with feathers twined in their long floating hair.
Thus through the sunset's golden gates, went out the autumn day.

Lo! in the night, the miracle of snow was wrought anew.
The gray-brown, arid plains were changed to marble pavements white;
Each rusty, rustling blade-like frosted fretted silver shone;
Each bush was turned to sculptured Indian, or deer in flight.
Autumn had vanished, and cold, ice-crown winter reigned alone;
And over all was spread a canopy of deepest blue.

AWAY FROM HOME.

BEAUTIFUL butterfly brown and white,
    With spots of black and gold,
Why are you here in the city's street;
    The city so sombre and old?

"The roses red and the roses white
    That climb on the granite wall,
To my clover field a message sent,
    And I came at their loving call."

POEM.

Read at a banquet of California Commandery, Loyal Legion, on the occasion of their twenty-first anniversary. At Mare Island Navy Yard, May 3d, 1892.

SING comrades, sing of peace
    Glad songs to-night;
Banished be grim war's face,
    Far from our sight.

Forget the cannon's roar,
    The sabre's flash,
The flag low in the dust,
    The rifle's crash.

Poem.

Forget the weary march,
  The bugle call;
Forget the empty sleeve,
  The prison wall.

Drink to the dear old flag,
  The stripes and stars;
Drink to the veterans brave,
  Covered with scars.

In silence, with bowed heads,
  Drink to the dead
Left on the battle field
  When all had fled.

Drink to a land at peace
  From shore to shore,
Heart to heart, hand to hand,
  Forevermore.

FOUND.

I WATCH the tender leaves, this April day unfolding,
    And look upon the shadows flitting o'er the lawn,
And I see children's faces, bright and winning—
    The faces of my darlings, long, long gone.

The first I see is Baby in his dimpled sweetness,
    Blue eyes, white face and little rings of curling hair;
I hold my hands out to embrace him fondly—
    Alas! they only meet the empty air.

Again I feel a chubby hand mine tightly holding,
    And guide two wee feet trying hard to cross the floor,
To see dear faithful Carlo soundly sleeping,
    In the warm sunshine just outside the door.

In sailor suit and hat, with many happy children,
    I see my schoolboy coming down the village street;
His hair wind-tossed, his glowing cheeks like roses—
    Again my schoolboy I shall never greet.

Away, away with all my sweetly tender dreaming;
    I hear a bounding step upon the oaken stair;
I look into the blue eyes bending o'er me—
    My baby, toddler, schoolboy, all are there.

THE CRYSTAL BELLS OF SANTA HELENA.

IN a garden upon a cliff
    High above the fair southern seas,
Hang crystal bells, with silver tongues,
    On branches of olive trees.

In the sunlight down on the beach,
    Play the tiny rippling waves,
And breakers dash against the rocks,
    And thunder in ocean caves.

But the winds are asleep the while,
    And the crystal bells idly hang,
As if out on the southern sea
    They never in melody rang.

When lo! from the bright golden west,
    Where the sky and the waters meet,
Came a breeze from a tropic land,
    And the bells rang out clear and sweet.

So many a heart that was mute
    Like the bells on the olive trees,
At the voice of love rings out clear
    As the bells at the southern breeze.

ANCESTOR MINE.

HE hangs upon the wall ancestor mine;
   No powdered wig, nor queue with ribbon tied.
No ruffled shirt, nor shoes with buckles wide,
No dangling sword, he wears, or feathers fine.
No knighted hero he of wars long past;
   He sits in tiny elbow chair of old,
   A little boy with hair of shining gold:
In dimpled hand a crimson whip holds fast,
A suit of mauve, with frills of dainty lace,
   Bright scarlet shoes, a brooch of jewels rare;
   His sweet young self looks out of ancient frame
With eyes of deepest blue—a soulful face;
A gentle mouth, yet firm, and face most fair;
   My great-great-grandfather, the wee one's name.

THE ASTRONOMER'S WIFE.

I WANT to thread the golden stars,
   A necklace bright to wear.
I want a diadem of stars,
   To rest upon my hair.

The Astronomer's Wife.

I want a dress from cobwebs spun,
    Flecked o'er with tiny stars.
I'd be a constellation new,
    Quite near to fiery Mars.

My hair, like flying meteor,
    Should float out into space;
The moon, like fan from far Japan,
    I'd hold before my face.

Then he I love, who now forgets,
    Would gaze on me each night;
He'd sweep the sky with telescope,
    Of me to catch a sight.

And there I would contented lie
    Up in the sky of blue,
Discovered first by him I love,
    A constellation new.

He then would think of me alway;
    Would give me his dear name;
I'd bring to my astronomer,
    Oh, joy! renown and fame.

THE PAGAN GIRL'S PRAYER TO THE SUN.

(B. C. 500)

O SUN, thou God who for ages my people
    Have worshiped, low in the sky, o'er the sea
There thou hangest, a red ball of fire,
    Tarry, oh tarry, and listen, I pray thee.

Thou who lightest up dark places with sunbeams,
    Thou who paintest the flowers and rainbows,
Thou who fillest with sunlight o'erflowing
    The cup of the lotus, list to my sorrows.

O bright sun, thou hast left me; thou hast fallen
    Down into the waves. Thy blood stains the sky
In the west, and lies red on the waters.
    Thou heardst not my sorrow, nor answered my cry.

A VIGNETTE.

LIKE Italian portrait by master's hand
    Or clear cut cameo, a face
That in my beautiful, ideal world,
    In my castle has a place.

WAITING FOR COLIN.

I AM growing old, my hair
    Once so golden, is now white like snow,
And I live in the far away past,
    The beautiful long ago.

Oft-times I stand at the door
    Of the farm-house, my earliest home;
The sun is sinking behind the hills;
    I wait for Colin to come.

Again I am little May,
    When I stand on the doorsteps so high,
The hollyhocks, covered with crimson flowers,
    Are half a head taller than I.

The wind the red clover sweeps,
    And the tinkling of bells I can hear,
The cows down the hillsides are coming now;
    I know that Colin is near.

    *   *   *   *   *   *

He was true to me 'till death.
    Now he dwells in the world of light.
I have been lonely for many years,
    But Colin seems near me to-night.

I wait for Colin alway.
  He will come when the sunset is bright;
Again I'll be his "own little May,"
  And golden my hair, not white.

UNDER THE ORANGE TREES.

THEY stood at the evening hour
  'Neath orange blooms sweet and white,
Beside a tropic sea,
  In the sunset's golden light.

He gave her orange blossoms,
  Oh! mockery in the thought;
She was bound by iron fetters;
  Their sweetness counted for naught.

The snowy, waxen blossoms,
  Nestling fondly side by side,
Should rest on other tresses,
  She could never be his bride.

THE OLD "HARTFORD."

SAILING out on the waters blue
   Of the San Francisco bay,
Under the flag they fought to save,
   On their anniversary day,

The heroes passed the old "Hartford,"
   Stripped of her warlike signs,
And to those loyal hearts there came
   Memories of olden times.

There was one among the number
   Brushed away a starting tear,
As he thought of Fort St. Philip
   And the fate that then seemed near.

From the ancient ship at anchor
   Sailors waved the Union Jack,
Fastened well to unused ramrod;
   Then for answer went floating back,

On the breeze the Nation's Anthem,
   Sweeping o'er the waters blue,
In honor of the old "Hartford,"
   From the veterans, tried and true.

SHADOW LAND.

INTO shadow land I wandered,
   Led by twilight's hand,
Gently from the sunset golden,
   Into that drear land.

Dusky shadows all about me
   Whispered sad and low,
Saying I should walk forever
   In their vale of woe.

Telling on my life forever
   Would their darkness stay,
As across the threat'ning heavens
   Then a dark cloud lay.

Half despairing wildly cried I
   To the sombre night:
"Take me from the gloomy shadows
   To the blessèd light."

Lo! the clouds were fringed with moonlight:
   Joy, O soul of mine!
There can never be dark shadows
   Save where light does shine.

WILD POPPIES.

THE STATE FLOWER OF CALIFORNIA.

BEAUTIFUL, golden wild poppies,
   That nod in the soft, balmy air,
Well were you chosen the emblem
   Of the land of all lands most fair.

Who planted you, golden poppies?
   Were you here when the world was new?
Were you painted by the morning?
   Do you mirror the sunset's hue?

Do you grow from seeds of bright gold
   That are hidden away from sight?
Are you stars come down from the sky
   That shine in the radiant light?

Are you golden cups o'erflowing
   With jewels of rain-drops and dew?
Why are you so constant-hearted
   To the State that has chosen you?

With gold you carpet the meadows,
   Like the gold paved "Land of the Blest,"
Wild poppies—the flower emblem
   Of the State of "The golden West."

THE KING'S RETURN.

IN MEMORY OF KALAKAUA.

ON the throbbing heart of the tropic sea,
    Like lilies, the fair islands lay,
Half asleep in the sun.
    The winds seemed to sing,
 "We wait for our king."

The spray, like numberless pearls, on the shore
Is cast by the generous hand
Of the blue southern sea.
    The waves seem to sing;
 "We wait for our king."

There are beautiful bridges of rainbows,
Fair nature's triumphal arches
Of sunbeams and spray drops.
    Sea-nymphs seem to sing;
 "We wait for our king."

Under the feathery cocoanut trees,
Shading eyes from the eastern sun,
Stand subjects most loyal—
    The birds seem to sing;
 "We wait for our king."

The King's Return.

In the fair island city flags flutter
Like tropical birds in the air,
And music is sounding;
   Each face seems to sing;
"We wait for our king."

In the heart of the queen in the palace
What rapture to welcome the loved
Once again to his home.
   What joy thus to sing;
"We wait for our king."

*   *   *   *   *   *

Far, far away out at sea is a sail,
Like the white wing of a wild bird,
On the bright golden sky—
   Air, earth and sea sing;
"We wait for our king."

The wing has changed to a bird, then a ship,
A grand man-of-war, on whose masts
Two nations' flags flutter.
   The ship that will bring
   The waited-for king.

Half-mast are the flags, draped in black the ship;
The sunbeams and rainbows are gone;
The waves wail and moan;
   The glad song has fled,
   The good king is dead.

"SOME ONE."

THERE'S something wanting in the morning,
    The city wears a sombre look to-day;
Song birds I'll tell the reason to you:
"Somebody" is away.

If I had wings, I would have followed,
    And sung my sweetest, tenderest songs, and gay;
I have not, and I am so lonely,
    For "Someone" is away.

The air is full of hope this morning,
    Birds never sang so sweet until to-day;
Not one fair flower had bloomed, I thought,
    Since "Someone" went away.

If I had wings, song-birds, I'd fold them;
    Here in the city I would rather stay.
I'll whisper low the reason to you;
    "Someone" comes home to-day.

CAST ASIDE.

A BABY sittting in the sunshine on the floor
   Tried with her dimpled hands to brush the
     sunbeams from her dress.
So sitting in life's sunshine we oft cast aside
   With thoughtless hands, counting as naught the
     brightness sent to bless.

MEMORIAL DAY.

IN a lonely spot beside the sea,
   'Neath sobbing pine trees, many, many miles away,
     Lies a soldier brave.
Like a pagan woman to the sun I cry:
     "Decorate his grave."

"O sun send down your beams most brightly;
Make on that grave, mourned by the ever restless sea,
     Blue violets grow.
O summer wild birds, sing o'er my soldier dead,
     A requiem low.

When on his grave, tributes of flowers
His soldier comrades brave shall place, they'll start at
    sight
        Of violets blue;
Nor dream, at prayer of mine, for love of him,
   .   The violets grew.

A LEGEND OF CALVARY.

RED-BREASTED robin airily poising
   On slender twig of an apple tree,
From far away land, and from long, long years past,
   You bring me a legend of Calvary.

Upon the cross our Savior was dying,
   A crown of thorns on His sacred head,
A little bird hovered pityingly near Him,
   When some who had loved Him, foresook him and
      fled.

The little brown bird from the cruel crown
   A piercing thorn took gently away,
And the crimson blood falling from that holy brow,
   Changed the sombre brown bird to a red breast gay.

A WHITE CHRYSANTHEMUM.

LAST night beside my hearthstone
    She sat in snowy dress,
The firelight touched her golden hair
    With many a fond caress.

She wore white autumn flowers,
    Like frozen stars they seemed;
One flower she left, else I should think
    Of angels I had dreamed.

THE HEART'S FIRELIGHT.

I SIT beside the hearthstone of your heart,
    A welcome guest.
I was a wanderer, without a home;
    You bade me rest.

I sang bright songs of hope, I was so glad
    You bade me stay.
I fanned the embers dull, and brighter grew
    The flame each day.

To be the chirping cricket on your hearth
    Is joy to me,
And you have promised that no other one
    Shall ever be,

The same that I am now to you, dear love,
    So close to thee.
No other shall e'er fill the corner bright
    You've given me.

DISCOVERY OF THE SUNSHINE MINE.

I HAD left the tired miners
    When the sun was turning to gold
The long line of purple mountains,
    And the tall peaks rugged and bold.

I was just a toiling miner,
    At work on the "Eagle's wing claim,"
Searching, alas! searching vainly,
    Yet hoping and toiling the same.

On my shoulder pick and shovel,
    That fair day in early June,
As I drew near our small cabin
    I was whistling a merry tune.

Discovery of the Sunshine Mine.

I gleefully called: "Come Sunshine."
   She was all in the world to me.
"Where are you hiding my Sunshine."
   Why where can my darling child be?"

The sunlight fell on the cabin,
   And danced in the open door,
A slanting pathway of glory
   It made on the rude wooden floor.

No answer, but silence—silence—
   Save the cry of a lonely bird,
And the summer breezes sighing
   Through the tree-tops was all I heard.

But where was my little daughter,
   My darling with bright golden hair?
"Where are you? Where are you Sunshine?"
   Then I cried in my wild despair.

For Sunshine, my little daughter,
   My one treasure, young and so fair,
Was always waiting to meet me;
   The thought of her drove away care.

In yesterday's fair June weather
   Up the cañon, rock-strewn and wide,
To find the first wild columbines,
   We had wandered at eventide.

Wild Poppies.

As swift as the bullet that flies
    From gun, to the heart of a deer,
As crushing, stunning, and hopeless,
    Came to me the terrible fear,

That Sunshine in search of flowers
    Up the trail had wandered away,
Then I, who had forgotten God,
    In my agony knelt to pray.

I thought of the icy cold winds
    From the peaks of eternal snow,
Of cruel, prowling, hungry wolves,
    And of chasms that yawned below.

Quick, half dazed and blind, I stumbled
    Up the cañon, wild with despair,
To search for my little Sunshine
    For my darling with golden hair.

Heart-broken, I wandered onward,
    I begged the sun longer to stay,
The night not to wrap its black arms
    Round the mountain's dangerous way.

Something bright gleamed just before me,
    Where the first wild columbines grew;
I hugged it close to my heart,
    'Twas a small, worn, copper-toed shoe.

Around a boulder I hastened,

A DREAM.

I DREAMED the chariot wheels of time had ceased
    to roll;
That the blue heavens were parted like a riven scroll;
That holy angels, with bright shining hair,
Floating about them in the summer air,
God's messengers from the heavenly land,
Had wandered down to earth from His right hand.
The sea gave up its dead from parted waves;
Like lilies fair, the dead forsook their graves.
My mother, radiant as evening star,
I saw, smiling upon me from afar.
I heard a voice of majesty that cried:
"Come all who love The Christ, The Crucified."
I hastened to the grave of one I love;
It was unchanged, the tall grass waved above,
And violets still threaded wreaths of blue,
And sunbeams turned to jewels drops of dew.
I whispered softly: "Wake love, come with me.
'Tis morning love, hasten, I wait for thee."
I threw myself upon his fast-sealed grave,
Above the heart I thought so good and brave;
I begged grim Death his iron chains to burst;
A voice proclaimed: "The dead in Christ, rise first."

MY MOTHER'S BIRTHDAY.

TO-DAY'S my mother's birthday, yet I cannot lay
    Fair flowers on her grave, it is so far away,
Nor with my face bent low among the daisies wild,
Whisper, "I love you mother, do you hear your
    child?"
And so alone I sit in revery to-night,
And wonder if earth's birthdays in that land of light
They keep, or count it life when through the pearly
    gate
They enter in the city paved with gold. I wait
An answer, but the night wind hurries silent by;
No answer comes to me from out the star-gemmed sky.

A WINTER'S DAY.

CALIFORNIA.

TO-DAY I hold pink rosebuds, lilies white,
    Dasies and wildwood violets in my hand;
  Dark ivy to the casement clings;
The sea a sapphire gleams, an emerald the land.
A tiny shadow, 'tis a tropic bird in flight,
    That cuts a sunbeam with its wings,
        Its scarlet wings,
        And glad song sings.

Such is fair California's winter day.
Where is the sparkling, dazzling, icy crown?
  The ermine robe on plain and hill?
The last year's robin's empty nest in branches brown?
The snow on trees? The little snowbirds? Flown
    away?
  The frozen lake? The moonlight still?
      The moonlight still
      On icy hill?

Where are the branches bending 'neath the snow?
The silver fringe of icicles upon the eaves?
  The marble of the hills and dells?
The north wind scattering far the dry brown leaves?
The frost upon the panes? The firelights's bright
    glow?
  The merry, merry sound of bells?
      The sound of bells
      Through icy dells.

Grim winter heard upon the mountains tall,
The softly wooing voice of the fair tropic sea.
  Felt kisses of the warm, sweet air,
The flower-filled air, that whispered, "Come with
    me."
Dropped ermine robe, let icy scepter fall,
  And stole from mountains down to land of all most
    fair;
      To land most fair,
      From icy air.

THE SEA IS A GRAVE TO-DAY.

THE sea is a grave to-day;
   On its bosom one young and fair,
Sleeps the long, long dreamless sleep,
   With seaweed twined in her hair.

Rocked by the billows she rests,
   And softly the winds o'er the deep
Sing of her who sleeps so well;
   "She will never wake to weep."

A sunbeam kissed her still face,
   And wrapped in the fleecy white spray,
She sank 'neath the waves she sought;
   And the sea is a grave to-day.

MORNING GLORIES.

MORNING glories climbing
  Upward to greet the dawning,
Sparkling with fair dewdrop jewels,
  Noon-tide and evening scorning.

Swayed by summer breezes,
  Kissed by droning, drowsy bees,
Lovingly, gracefully clinging
  To branches of stately trees.

Coquetry's the emblem
  That has been chosen for you;
But ne'er to radiant morning
  Were ever flowers so true.

For you fold your blossoms
  From the noonday sun away
And have no thought of aught of earth
  'Till dawn of another day.

ONE WILD MARCH NIGHT.

ONE wild March night when the wind was high
    Before the fire sat Dora and I.

Grim was the fireplace, deep and wide,
Two tall black andirons stood side by side.

Stories of goblins and elves I told
'Till the maple logs turned to living gold.

I said to Dora: "If some tiny elf
Should say you no longer could be yourself,

Pray who would you be, my love, my life?"
She answered: "I'd be Dick, your second wife."

A PRINCE.

J. N. E. W.

IN his home in the valley, that afar
  Like a dream of beauty, wanders away;
With background of mountains that kiss the sky,
  I saw a young Prince to-day.

The Prince was enthroned in his mother's arms,
  His beautiful Queen, so winsome and fair,
A sunbeam stole through the window, and placed,
  A crown on the baby's hair.

His cheeks had the tint of the pink sea shell
  And his eyes the look of the coming king.
Oh! I wonder did others see the crown
  I saw, on the Prince I sing?

AWAY.

THE foils are idly crossed upon the wall,
  Tied with a silken ribbon, soft and wide,
The color that his lady wears, fair blue;
  Shakespeare much read, alas! lies tossed aside.

Wild Violets.

I am the lady who the fair blue wears;
  I am his heroine in Shakespeare's plays;
Often I've wielded one bright steely foil;
  Alone I dream away the autumn days.

Out from our home my hero brother's gone,
  Out to win bread, perhaps renown and fame;
Life that was like one long bright summer day,
  Never again can be to me the same.

WILD VIOLETS.

BECAUSE you mirror the sky
  In colors of heaven's own blue,
For your sweet and dainty selves,
  Violets, I love you.

For thoughts of your forest home,
  Its wild flowers sparkling with dew;
For the sake of the giver kind,
  Violets, I love you.

THE ROYAL SUCCESSION.

SUMMER had lingered long on the plains,
   Summer robbed of her beautiful green;
Heart-shaped leaves of the cottonwood trees,
   Motionless waited the autumn wind keen.

Dust and ashes the gray-brown earth seemed;
   Birds had flown southward to find fresh flowers,
Autumn stood tiptoe on mountains cool,
   For summer's reign was counted by hours.

Passionate summer shed great burning tears
   And turned the sky to a huge black cave,
Where fiery lightning serpents played;
   Soon dry leaves showered on summer's grave.

Lo! in the morning fair autumn reigned,
   An eastern queen dressed in colors bright,
From mountain tops like a goddess fair,
   She came to the plains in the soft moonlight.

ON THE BEACH.

THE white-crested waves at my feet
   Tossed a piece of a ship lost at sea;
I seized it quick with my trembling hands,
   Then I tossed it away from me.

In fancy I saw a proud ship,
   Homeward bound from the bright sunset land,
And naught was left of that white winged bark,
   But the fragment tossed on the sand.

No avail to cast it away,
   For great waves brought it back to the strand,
As memory brings all our shipwrecked hopes,
   To us with a pitiless hand.

A VALENTINE.

BLUE VIOLETS EMBLEM OF LOVE.
—Language of Flowers.

LIKE a quiver of arrows my thoughts—
　　Some are golden, some silver, some steel,
Alone to-day with fair Cupid's bow
In my high eastern window I kneel.
I wreathe one arrow with violets blue,
Then I bend the bow and it flies to you.

NEW YEAR FANCIES.

FORGETTING the past, with its dreams
　　That faded away
Like the radiant dazzling colors of sunset
　　That came not to stay.

The fleecy white clouds, you fancied
　　Were castles most fair
With towers and turrets, with banners of sunbeams
　　Afloat in the air.

Forgetting the past, with its dreams
  Like tales that are told,
Dream dreams brighter, aye fairer than ever before
  In years now grown old.

EITHER WAY.

BLUE Cloud, an Indian bad,
    Paused long before his gate;
He had been drinking whisky,
    And stayed out rather late.

Blue Cloud was always bad,
    To-night he longed to fight;
Alas! poor little squaw
    Asleep in the moonlight.

" If she has gone to sleep
    I'll beat her black and blue;
If she's up burning wood,
    Why then I'll beat her too.

But if the room is cold
    I'll beat her." Blue Cloud said;
" Or if she watched not for me
    I'll beat her sleepy head."

Poor little dusky squaw,
  Though dutiful you be,
You surely will be beaten.
  White men do you see?

MY NEW ENGLAND HOME.

A VISION fair of a quiet town
  Memory brings to me to-night;
A town on the banks of a river chill,
  Asleep in the pale moonlight.

Tall trees stand on the river banks
  Mirrored ghostly in depths below;
Green tangled wealth of blackberry vines,
  And golden-rod, by the roadside grow.

Across the village street the elms
  Whisper together in voices low,
And moonlight soft in silvery showers .
  On the brown earth falls like snow.

I see the white church on the hill
  And the clock in its tall tower,
With its iron hands together clasped,
  As it tolls the midnight hour.

The moonlight is fading fast away,
  My home is now by a tropic sea;
Outside my window are stately palms;
  But my childhood's home is dear to me.

THE LOVELIEST PICTURE.

(MABEL M———)

IN an artist's studio I looked
  Upon many pictures, grand and bold,
On purple mountains crowned with snow,
  And radiant sunsets of crimson and gold.

I stood almost entranced before
  A fairy-like, far New England scene;
A brown road leading, leading away
  Through heart of a forest robed in summer's green.

I felt the cool moist air of the woods
  And I heard a wild bird's mournful cry,
I saw starry blossoms, nodding ferns,
  And could hear a tiny brook murmuring by.

I turned from painted canvas, and lo!
  A lovely, living picture was there,
A little maiden, only just seven,
  Gracefully poised, and with sweet childish air.

From under her wide felt hat she gave
    Coquettish glances from sparkling eyes,
Like wild pink roses her dimpled cheeks,
    Her eyes were the color of soft azure skies.

"Life's Morning," seemed her beautiful face,
    On which rested no shadow of care,
Each canvas showed a master's touch,
    But I thought her the loveliest picture there.

A PRAYER.

CHRIST pity all sailors to-night
    On the tempest-tossed sea.
Say "peace" to the storm,
    The waves obey Thee.

I hear the sea lash the great rocks.
    Stars are hidden from sight,
The winds wail and moan;
    Christ keep all to-night.

My heart bird-like flies to a ship
    Far away out at sea;
Oh pity! and bring
    My sailor to me,

Or out on the wings of the storm
  Send my soul to his side,
Forever to be
  In heaven his bride.

IN THE GLOAMING.

I SAW you in the gloaming,
  When, wrapped in silver mist, the city
Like a fair bride stood in fleecy veil.
No sun, no stars, only the cold grey fog;
Even the winds had ceased to sob and wail.
Now you are real to me,
While I am still, and ever must be,
Like the cold mist, silvery white,
That melts away so soon at the sun's kiss—
That ghost-like glides away at morning's light.

COMPENSATION.

DARK clouds rolled over the sky,
    And but one star could I see;
I cried in my wild despair:
    "Let the bright star shine for me."

But the purple clouds rolled on
    And hid the star from my sight,
When lo! where the clouds had been
    The fair moon was shining bright.

A CIGARETTE.

THE day is dying. In the western sky
    The sun still lingers, brightness lies on waves,
The fallen shield of day. There comes to me
A vision fair, as curling mist-wreaths fly
Across the sun like puffs of smoke. There lies
    Upon the window-sill a cigarette.
    A tiny thing from Egypt far, and yet,
The lotus floating on the Nile, blue skies,

Tall palms, and faces dark, fade fast away;
  And Venice rises up from waves of blue,
  Its waters tinted with the sunset's hue;
And melody of bells at close of day.
A traveler.—The sunset lingers yet
As does the vision, and—the cigarette.

WAITING AT THE GATE.

THE birds are singing sweet vespers,
  As I stand by our cottage gate;
In the glory of slanting sunbeams,
  I watch for my loved one and wait.

The city across the waters
  Seems fading into the sea,
As I watch a boat coming, coming,
  That's bringing my loved one to me.

I often think in the sunset,
  As among the flowers I wait,
And the birds are singing sweet vespers,
  Shall I stand at the pearly gate?

Shall I stand in untold glory?
  Shall I watch a boat stem the tide?
Shall I welcome, as now, my loved one
  To our home on the other side?

IN THE CATHEDRAL.

OH where is she now I wonder,
   The girl with the pale golden hair,
And sweet white face, and violet eyes,
   Who knelt in the church at prayer?

Have the soft Italian breezes
   Kissed the roses back to her face?
Do her eyes have still the saintly look
   That they wore in that holy place?

Oh where is she now I wonder
   The girl with the pale golden hair?
In her English home?—In Italy bright?
   Or in heaven an angel fair?

SAN JUAN BY THE SEA.

I SAW thee in the sunset,
    Fair San Juan by the Sea,
Like a golden band of glory
    Looked the western sky to me.
The deep blue of the waters
    Met the orange of the sky,
That melted into palest gold
    Where one star shone out on high.

BEFORE THE HOLIDAYS.

IN our far off New England home,
    At the side of the chimney wide,
Ever on Christmas eve I used to hang,
    Maxy's small stockings side by side.
    Now Maxy's away at school
        In a university town,
    To-morrow is Christmas day,
        And the snow comes drifting down.

If I had but one golden coin,
   But a crumb to a millionaire,
'Twould give me the sound of my darling's voice,
   A glimpse of his brown curly hair.
      Now Maxy's away at school
         In a university town,
      To-morrow is Christmas day,
         And the snow comes drifting down.

Upon one merry Christmas eve
   Maxy mine made a boyish boast;
"Some Christmas mamma, my stockings I'll fill
And bring you what you love most."
      Now Maxy's away at school
         In a university town,
      To-morrow is Christmas day,
         And the snow comes drifting down.

    *   *   *   *   *   *

It is Maxy's step on the stair,
   Oh! joy it was no idle boast;
"Some Christmas mamma, my stockings I'll fill
And bring you what you love most."
      Now Maxy is not at school
         In a university town,
      To-morrow is Christmas day,
         And the snow comes dancing down.

AFTER THE HOLIDAYS.

I WATCH this cold, bright winter's day, the sun-
   beams dancing
Like flocks of yellow birds across the floor,
And listen for the bounding step of Maxy;
   Alas! I know the holidays are o'er.

I marvel much at some fair Spartan mother sending
   Her noble, loving boy to far-off battle-field,
Smiling, as with untrembling hand she buckles
   Over her darling's heart a silver shield.

I wonder if she ever roamed the meadows holding
   A small brown hand in her's, searching for daisies
      white,
And buttercups, like fallen stars from heaven,
   Some summer morning, bathed in rosy light.

I wonder if beside some marble fane, sad weeping,
   Mother and son have mourned their Spartan soldier
      dead;
Have sweet white flowers placed, and laurel wreaths,
   And broken prayer in holy temple said.

There still are many like the Spartan mother sending
　Into life's battle-field, their boy, their joy and pride,
With smiling face, but aching heart; and praying
　"The God of battles" to be on his side.

WHITHER.

IN my window an empty cage,
　　The bird has flown, who can tell where?
Is it stranger the soul has gone
　　And a cold form is lying there?

SUSPENSE.

THE sky and the sea like two nuns
　　Wear mantles of gray,
And like a black cross seem the masts
　　And the yards of a ship far away.

Is it coming, coming to me
　　This heavy black cross?
Shall the hopes and the joys of my life
　　Suffer pitiful shipwreck and loss?

Safe.

The ship like a bird on the wing,
  Seems only to stay.
Alas! it is coming, it tacks,
  Oh! thank God it is sailing away.

SAFE.

AT the ebb of the tide, a stately ship,
  Sailed away to a southern coast;
In the moonlight pale, with sails unfurled,
  It seemed but a white, sheeted ghost.

On the midnight tide, it drifted away;
  Far away on the trackless main;
The stars shone bright, but the night-wind wailed,
  "It will never come back again."

* * * * * *

The ship came back from the sunny south coast,
  Like a bird, with its white wings spread;
The morning sun made the sea like gold,
  And the wind with its warning had fled.

SUNSET FANCIES.

AT THE GOLDEN GATE.

FLAME-COLOR, orange and palest gold,
    Sunset stairs to the azure sky,
Up which the summer day has gone,
    Trailing her robes of amethyst dye.

Shadowy grows the stairway, then dim,
    As night in somber robe comes down,
Her dusky mantle gemmed with stars,
    On her forehead a crescent crown.

WHERE?

IN my tiny boat alone,
    Just inside the Golden Gate,
From tropic shore, or from out at sea,
    For message to come I wait.

'Twas sunset an hour ago
    And long slanting lines of light
Closed the way to the ever restless sea,
    Through the golden gateway bright.

But the hand of twilight came
  And loosened the yellow bars,
Now a silver pathway across the waves
  Is lighted by gleaming stars.

O city upon the hills,
  A queen rising out of the sea,
Your thousand firefly lights seem to call;
  "Come back, we've a home for thee."

I hear; but "The sea is His;"
  If He calls me I must go
Out on eternity's fathomless sea;
  What He wills is best, I know.

TO ADA REHAN'S PICTURE.

UPON the city's street,
    I paused at vision fair;
Eyes where genius shines,
  Wealth of waving hair;
Snowy neck and arms,
  Mouth like Cupid's bow,
Dream of poet's soul,
  Dress of long ago.

Form of faultless mould,
  Poise of stately grace;
Every day I gaze
  On that perfect face;
And I turn away
  With a fond regret,
Though soon far from me,
  I can ne'er forget.

ALPINE BARRY.

HERO, MARTYR.

LOFTY Alps lifting up to the sky
  Giant helmets and nodding plumes white,
Sea of ice stretching far, far away;
  Sea of fire in sunset's red light.

Holy monks out into the gloaming
  Sending brave Barry, rescuer, guide;
"Jesu protect all lost ones," they pray,
  "Bring each again to his fireside."

Avalanche sweeping with awful sound,
  From the tall peaks to chasms below;
Out from the light of the hospice door,
  Out on the white waste, the trackless snow.

Alpine Barry.

Brave Barry going, dog that had saved
   Forty lost ones from the Alpine cold,
Hardy travelers of many lands,
   And one a fair boy with curls of gold.
*   *   *   *   *   *   *
A soldier struggling up the wild pass,
   Fighting the fierce storm that sweeps the land,
A soldier lying beneath the snow
   With his trusty sword clasped in his hand.

A cry of despair from stift'ning lips;
   Brave Barry hast'ning a life to save,
Through the blinding storm and cruel snow
   Finding his way to the soldier's grave.
*   *   *   *   *   *
A soldier wan before the fire
   Telling the monks in its cheerful glow,
Of his dreadful battle with the storm,
   And his grave in the white drifted snow.

Of a savage beast with warm moist breath,
   Of gleaming eyes that above him bent,
And that the sword he grasped in his hand
   Through the heart of the monster he sent.

A look of horror upon each face;
   An aged monk in a low voice said:
"Oh! brothers it was our noble dog."
   In the dawn they found brave Barry—dead.

TO————.

WOULD the sun shine as bright as now
    Dear heart if you were gone?
Would birds upon the trees
    Forget their song?
Would flowers bloom?
    Would soft winds whisper to the sea?
Would hearts be merry, light and gay?
    Could such things be?

I know the sun would shine as bright
    Dear heart if you were gone.
The happy birds would not
    Forget their song.
Flowers would bloom
    Soft winds would whisper to the sea.
To many, life would be as sweet,
    But not to me.

THE SAPPHIRE SEA.

THE sky is a sapphire sea;
    The stars so sparkling and bright,
Have caught and reflected the glory
    Of "the city which hath no night."

Blue, blue is the sea and at rest,
    Save where sky and mountains meet;
There long white fleecy lines of clouds
    Like surf on the hill-tops beat.

And lo! there's a crescent-shaped boat
    Of silver, upon the sea;
And in it are jewel-crowned ones,
    Who waft a message to me.

For I am a mermaid glad;
    I dwell under the sapphire sea,
I gather bright jewels and pearls,
    The work that is given me.

A PARIS BONNET.

DEACON Smallman to the city
    Business called, one bright spring day.
"Bring me home a lovely bonnet,"
    Said his young wife, pretty May.

She was quite a living picture,
    Gypsy-faced, and full of life.
"Too worldly minded" gossips said,
    "For a sober Deacon's wife."

At the milliner's the Deacon
    Heaved a regretful sigh;
True the bonnets were "reel putty,"
    But the prices were so high.

At last the charming milliner
    Said: "Here's a Paris poke
Only six bits," and then she coughed
    'Till the Deacon thought she'd choke.

"'Twill be so very sweet," she said,
    "Trimmed with buttercups, you know
And poppies, and—and clover leaves
    And with just a tiny bow."

A Paris Bonnet.

Well, the Deacon bought the bonnet,
   And May's rougish gypsy face
Under the stylish "Paris poke,"
   Was the envy of the place.

One day through quiet Meadowtown
   Marching down the village street,
Came the Salvation Army,
   With much music loud, not sweet.

Oh clouds! blot out the sunshine fair.
   Each Salvation woman wore
"A Paris poke," minus flowers.
   Alas! the good Deacon swore.

"This world is but a stage," we know,
   "Men are actors," so they say.
At Deacon Smallman's rural home
   There was held a matinee.

ALL THAT REMAINS.

IN a fair southern land, an old church stands
    A ruin, with curious roof of tiles;
Through crumbling arches gray, star tapers gleam,
    And moonlight shadows wander up its aisles.

Through rifts in broken roof, sunbeams caress
    The pictured face of saint with golden hair;
Time's hand has blotted out each one save her's,
    Of all the holy faces gathered there.

When noble lord, and peasant too, pass by
    That ancient church upon their sev'ral ways,
Before the saint with the bright golden hair,
    In loving homage each one kneels and prays.

Like that old Spanish church, many a life
    A ruin now, once was a holy place;
Upon whose walls of memory still hang,
    The picture of some loving, saintly face.

THE SUN HAS GONE DOWN.

SUNSHINE over the city,
    And sunlight upon the bay,
Peace and hope, joy and gladness,
    Life but a bright summer's day.

Fog, and mist, and the darkness,
    Over the sea and the town;
Houses and ships are spectres,
    For oh! the sun has gone down.

Life was to me all sunshine,
    When out on the shoreless sea
Sailed one I loved, and now
    The sun has gone down for me.

DO THEY KNOW?

DO the loved dead know, in their bright heavenly
    home,
  When on their dreamless beds are laid earth's
    flowers sweet,
When blue forget-me-nots, and lilies white
  Upon their lonely graves the wild-flowers meet?

It were not strange if earthly flower-full hands,
  And angel hands should bridge death's river, dark
    and wide;
Or if our Father, earth's fair, fading flowers,
  Should make immortal on the heavenly side.

CALIFORNIA.

NEVER a land so fair,
    Land of sunsets golden,
Kissed by a sapphire sea,
    Land of Missions olden.

Home of birds and flowers,
    Olive and tropic palm;
Black-winged storms sweep never
    O'er its summer skies calm.

Land of gold and silver,
    Land of honey and bees,
Land of wine and plenty,
    Land of the giant trees.

Of California
    Proud may her sons well be,
Proud of a land so fair,
    Kissed by a sapphire sea.

MY WATCH.

YOU are dying my little watch,
    Your heart is beating slow;
I hold you in my hands,
    I listen,—oh! so low
Is your voice that used to chirp
    Like a cricket, long ago.

Little watch, I grieve, oh I grieve
    That your life's work is done;
That your heart will not beat
    At setting of the sun,
That your hands will be at rest,
    That your race is almost run.

You were old, O my little watch,
    When first into your face
I looked with childish eyes
    Before the fire place;
The red light dancing gaily
    On your tiny jeweled case.

My Watch.

My loving gaze followed your hands
    As in the rosy glow
I sat near grandmama,
    Grandma who loved me so.
She promised you should be mine
    When a lady I should grow.

Little watch she left me alone
    With none to care for me,
Lonely and sad of heart
    I had a friend in thee.
A talisman since childhood
    You have ever seemed to me.

Like a fair Egyptian princess
    Of ages now grown old,
Little watch I'll keep thee
    In thy small case of gold,
Wrapped like ancient mummy
    In many a silken fold.

NIGHT AT SEA.

THE stars like tapers burn
    Across the waters deep;
The winds, like summer breezes sigh,
    In peace I'll fall asleep.

My Father lit the stars,
    He stilled the storm-tossed deep,
His voice controlled the winds;
    Therefore in peace I'll sleep.

A LAUREL WREATH.

THE laurel trees wandered down to the shore
    To mirror their faces in the blue waves.
The summer breeze whispered gently to them
    Of sea-nymphs who dwell in pearl-strewn caves.

The moonlight lay like a silvery shield
    With moving laurel leaves traced on its side,
From out of the ocean Neptune came,
    To choose a crown for his sea-nymph bride.

He gathered a wreath of bright laurel leaves;
And sailors oft see in the moonlight fair;
In Nautilus boat, the Ocean Queen,
With a laurel wreath on her waving hair.

DAY DREAMS.

THE countless stars are dreams we dream when
    we're awake,
But ev'ry morn the golden sun blots out the stars,
Or night with black cloud-curtains shuts them from
    our sight;
Yet when the sun and clouds are gone, and in the sky
The moon, a silver crescent crowns the evening hour,
They come again, and yet again, night after night.
We call dreams "toys," because we may not keep
    them now;
But when we walk among the stars, as angels do,
Perchance we'll find them real, not "toys," our day
    dreams bright.

ETERNAL SILENCE.

DEAD in your coffin lying,
    Cold lips of ashen hue,
Brow of marble as peaceful
    As a cloudless sky of blue.

Lips oh! so cold and ashen,
    You never move to tell
If your spirit eyes have opened
    To light of heaven or hell.

White lips that once were ruby,
    Death's secret so well you keep
That the living heart misgives
    Lest you sleep the endless sleep.

TO-MORROW WILL BE BRIGHT.

THE sea to-day is sad
   It wears a mantle of gray,
And ships are but shadows dim
   That were white-winged yesterday.

To-day the rain-drops fall,
   And winds have a sullen roar,
But to-morrow the sun will shine
   As bright as ever before.

The ships, now phantom barques,
   Will gleam in the glad sunlight;
Heart of mine so sad rejoice,
   For to-morrow will be bright.

UNDER THE SENTENCE OF DEATH.

UNDER the sentence of death,
    A prisoner in his cell;
Like a string of beads his days,
    And he knows their number well.

Under the sentence of death,
    All who walk life's way;
None but the Judge knows the hour,
    Only He the fatal day.

A KING'S DAUGHTER.

IF upon the city's street
    My fair Princess you should meet,
  Ina, with her gentle face so fair,
In her simple woolen dress,
You would never, never guess,
  To a royal kingdom she was heir.

A King's Daughter.

Often those who know her well
To each other softly tell
   Of her life, so quiet, yet so grand;
That upon her golden hair
Rests a crown of jewels rare,
   Placed there by her loving Father's hand.

Tiny cross my Princess wears,
As a token that she shares
   Burdens with all children of the King.
Like the North Star shining bright,
Sea-tossed ones she guides aright
   To the sure, safe shadow of His wing.

Earthly kingdoms are laid low,
But her Father's throne we know
   Through eternity shall surely stand;
Here, she has many a care,
But she'll reign forever there
   A fair Princess, at the King's right hand.

CHANGED.

THE south wind whispered in merrie May,
  "Come, come quickly, flowers fair;"
And dainty blossoms pink and white
  Covered the apple trees brown and bare.

Gay dandelions in meadows gleamed,
  The grass showed many soldier-like blades,
The maiden's-hair nodded, and violets blue
  Nestled close to the trees in sylvan glades.

The bees buzzed about among the flowers
  With a cheerful, cheering, constant sound,
And the little bird sang its soul away
  To the fond loving heart it had found.

But the golden dandelions now
  Are fluffy bits of browny fuzz,
And the bees that kissed the flowers fair
  Have lost their cheerful, cheery buzz.

In the hearts of yellow roses they
  Drone a dreamy, drowsy tune
All about honey, honey sweet,
  In the mid-day hours of June.

The birds have lost their sweet love notes
   And sing to fledgings a lullaby;
And oft-times clouds like black-winged birds,
   Sweep across the soft blue sky.

June has roses and rainbows too
   And many a perfect summer's day,
But for fairy-like, fragile beauty fair,
   There is never a month like May.

FAREWELL.

TO-DAY I have put on a snowy gown,
   And fastened a white rose upon my breast,
As if 'neath a coffin lid I must lie,
   In the long, long dreamless rest.

For to-day you'll look your last on my face,
   And perhaps your eyes will be filled with tears,
Because I'll be dead to you, oh my love!
   Though each may live many years.

Then think of me ever in snowy gown,
   With one white rose just over my heart;
There, kiss me farewell dear, I love you so,
   Just one kiss:—and then we part.

PICTURES ON THE WALL.

SUNLIGHT and shadows only you see;
    You say the walls of my room are bare;
To you they are only cold white walls,
    To me they are covered with pictures rare.

A MARINE VIEW.

The moon coming up from out of the sea,
    Making a pathway of pale golden light,
Across the blue waves from the distant sky,
    A fluttering sail, like a bird's wing white.

A young girl watching, watching the sail,
    Watching the boat cross the pathway of light,
Her brown hair tossed by the summer breeze,
    Watching the white sail drift out of her sight.

A FLOWER PIECE.

Clover blossoms, red and white,
    Dandelions and buttercups too,
Sweetbrier roses, a wide-brimmed hat,
    Twined with a ribbon of faded blue.

A COMPANION PIECE.

Sweet pond-lilies out of the water
  Holding their faces, gentle and fair,
Cat-tails nodding, and brown rocks covered
  With tender mosses and maiden's hair.

A NEW ENGLAND LANDSCAPE.

The misty light of Indian summer,
  Soft'ning the brown of a farm-house old,
Cornfield and meadow, and slanting sunbeams
  Turning the leaves of the maples to gold.

A MAY MORNING.

An apple tree covered
  With blossoms pink and white,
Bees and butterflies coqueting,
  Bathed in the morning's light.

ON THE PLAINS.

Lonely plains stretching away to the west, —
  Sage brush and prickly pear;
White covered wagons toiling
  Slow in the hot, noontide glare.

ALONE.

A pine tree among the rocks,
  High up on the mountain's crest,
Defying the bolts of Heaven,
  In its branches an eagle's nest.

A ROCKY MOUNTAIN LANDSCAPE.

Snow-crowned giant peaks to heaven uprising,
    A cascade dashing down a cañon deep and wide,
A lonely cabin, like an eagle's nest,
    Perched on an o'erhanging ledge of mountain's side.

A SOLDIER'S GRAVE.

A soldier's grave, o'er which Mt. Shasta like a sentinal, keeps guard;
    A soldier's lonely grave, where God's own hand has planted flowers white;
A comrade, faithful, unforgetting, standing by that grave alone,
Save for "a wide eyed rabbit" looking on in wonder, undismayed.
    A background radiant, the golden glory of the sunset bright.

Painted by fancy in lonely hours,
    Memory's pictures though they be,
Europe's palaces never held
    Pictures more life-like and real to me.

BESIDE THE SEA.

ALL the sunbeams of the sky seem dancing
    On the sparkling tropic sea,
And the great waves ceaseless moan and thunder
    In their solemn majesty.

But across the sky, like birds quick passing,
    Shadows fall upon the waves,
As if golden sunbeams danced too gaily
    Over sailors lonely graves.

Thus across the brightness of life's pathway,
    Sorrow comes alas! to all,
As upon the sparkling tropic ocean,
    Dreary, dusky shadows fall.

A GOLDEN PATHWAY.

I DREAMED that I stood upon the edge
    Of a river deep, and chill and wide,
In the twofold gloom of night and clouds,
    And I *must* cross to the other side.

I heard no sound of swift coming oar,
  I saw no sail like a bird's wing white,
The stars were blotted out by the fog,
  "The City" was hidden from my sight.

When lo! through "severing clouds" the moon
  A pathway made to the other side,
And one I loved was waiting for me
  As fearless I crossed the river wide.

NEW YEAR'S EVE.

THE endless years are only beads
  Strung on the threads of time,
And some are bright like golden ones,
  And some like amber clear,
While others seem like molten lead,
  And dimmed by many a tear.

To-night I held a shining bead
  And with reluctant hand
I grasped the new, and like a nun,
  O'er it I said a prayer;
If golden bright, or inky dark,
  I begged the Father's care.

MY TRAVELER.

GOD keep all who travel to-night
  By sea or by land;
Father in heaven hold them
  Close with thy powerful hand.
Keep them, O Father, from danger,
  Danger by land and sea,
Safe for those who love them;
  This is my prayer to Thee.

EVERY MORNING.

FROM open casement she waves her hand
  And follows me with her eyes of blue
And smiles on me as I leave each day,
  Aye, sweet as the angels do.

Some way on the crowded city's street
  And 'mid whirl and strife for wealth and fame
She seems to be near, my guiding star,
  Smiling on me just the same,

As from the window where roses climb,
   She wafts a good-bye to me each day;
It is joy to work for wealth and fame
   At my darling's feet to lay.

JEWELS FROM UNDER THE SEA.

FANCHON stood by the blue summer sea,
   Fanchon who came from a foreign land;
Sea-nymphs she saw in each crested wave,
   Sparkling jewels in each sea-kissed hand,
     Jewels from under the sea.

Fanchon held out her beautiful hands,
   Fanchon, whose hair is like fine-spun gold;
Called to the sea-nymphs in sweetest tones:
   "Bring me the gleaming jewels you hold.
     Jewels from under the sea."

The sea-nymphs' came on the great green waves
   Which like death shut her out from our sight,
When lo! in the sunshine Fanchon stood,
   Sparkling and gleaming with jewels bright;
     Jewels from under the sea.

TOO SOON.

THE moon rides like a silver boat to-night
   Upon the clouds, white-crested, sky-sea waves;
From solemn pine an eagle wings its flight
   To lofty crags, and peaks, and lonely caves.
Through bare, brown branches of the forest trees
   The wind, with voice of Indians of long ago,
Wails down the cañon, then, like summer breeze,
   Whispers to hardy mountain flowers low.
A timid deer, down to a lake so clear
   It mirrors a bright star that shines on high,
Comes down the trail, strewn with leaves sere and brown
   To drink under the star-gemmed sky.

    *   *   *   *   *   *

The clouds have blotted out the crescent moon
   And the bright stars in sky and lake of blue,
As light is blotted out of life too soon
   By hands we trusted and believed were true.

PLATONIC FRIENDSHIP.

MARBLE maiden fashioned in wise Plato's school'
Sculptured fair with wondrous classic art,
Crowned with laurel wreaths unfading,
   Yet she had a human heart.

She was but a scholar dull in that great school;
   Though at first she grasped the "pure ideal,"
Glances from dark eyes soon taught her,
   That earthly love is real.

One small marble hand in friendship she extended,
   While the other pressed her throbbing heart,
Nature, woman taught to worship;
   Plato teaches classic art.

UNDER A MIMOSA TREE.

THE dewdrops hang on the bending grass,
    A dragon-fly cuts a sunbeam through,
The moaning Cypress trees lift sombre arms
    Up to skies of cloudless blue.
A humming-bird sips from golden cup,
    In the hedges a hidden bird sings,
And a butterfly among the flowers
    Tells me my soul has wings.

TWO STARS.

A BLUE lake among the hills
    With a fringe of shadowy pines,
Above a glorious star
    That sparkles and gleams and shines.

A star in the clear blue lake
    That smiles to a star above,
The type of a human heart
    That mirrors the Father's love.

THE CLOCK ON THE TOWER.

UPON the city's crowded street
    There stands a tall stone tower,
And up almost among the clouds,
    A clock proclaims the hour.

It is a mentor, true and good,
    To all, who hurrying by
Consult its placid face, for Oh!
    "Figures can never lie."

That "life is passing quickly by"
    Unto all it tells alike;
To discontented workingmen
    It says; "I never strike."

One day I took an untried street,
    (Alas! confidence misplaced)
I found the clock I trusted well,
    Like man, was many faced.

www.ingramcontent.com/pod-product-compliance
Lightning Source LLC
Chambersburg PA
CBHW031401160426
43196CB00007B/849